Introduction to Blockchain Technology

T0074174

Published 2023 by River Publishers

River Publishers

Alsbjergvej 10, 9260 Gistrup, Denmark

www.riverpublishers.com

Distributed exclusively by Routledge

605 Third Avenue, New York, NY 10017, USA

4 Park Square, Milton Park, Abingdon, Oxon OX14 4RN

Introduction to Blockchain Technology / by Ahmed Banafa.

Routledge is an imprint of the Taylor & Francis Group, an informa business

ISBN 978-87-7022-160-3 (print)

ISBN 978-10-0092-267-7 (online)

ISBN 978-1-003-42626-4 (ebook master)

A Publication in the River Publishers series
RAPIDS SERIES IN COMPUTING AND INFORMATION SCIENCE AND
TECHNOLOGY

While every effort is made to provide dependable information, the publisher, authors, and editors cannot be held responsible for any errors or omissions.

Introduction to Blockchain Technology

Ahmed Banafa

Professor of Engineering at San Jose State University USA
Instructor of Continuing Studies at Stanford University USA

NEW YORK AND LONDON

Contents

Preface

I dedicate this book to my late father.

Blockchain is an emerging technology that can radically improve transactions security at banking, supply chain, and other transaction networks. It's estimated that Blockchain will generate $3.1 trillion in new business value by 2030. Essentially, it provides the basis for a dynamic distributed ledger that can be applied to save time when recording transactions between parties, remove costs associated with intermediaries, and reduce the risk of fraud and tampering. This book explores the fundamentals and applications of Blockchain technology. Readers will learn about the decentralized peer-to-peer network, distributed ledger, and the trust model that defines Blockchain technology. They will also be introduced to the basic components of Blockchain (transaction, block, block header, and the chain), its operations (hashing, verification, validation, and consensus model), underlying algorithms, and essentials of trust (hard fork and soft fork). Private and public Blockchain networks similar to Bitcoin and Ethereum will be introduced, as will concepts of smart contracts, proof of work and proof of stack, and cryptocurrency.

Readers will understand the inner workings and applications of this disruptive technology and its potential impact on all aspects of the business world and society.

This is book is for everyone who would like to have a good understanding of Blockchain Technology and its applications and its relationship with business operations including: C-Suite executives, IT managers, marketing and sales people, lawyers, product and project managers, business specialists, students. It's not for programmers who are looking for codes or exercises on the different platforms of Blockchain.

Acknowledgment

I am grateful for all the support I received from my family during the stages of writing this book.

About the Author

Prof. Ahmed Banafa has extensive experience in research, operations and management, with a focus on IoT, Blockchain, Cybersecurity and AI. He is the recipient of Certificate of Honor from the City and County of San Francisco, Author & Artist Award 2019 of San Jose State University. He was named as No. 1 tech voice to follow, technology fortune teller and influencer by LinkedIn in 2018 by LinkedIn, his research featured on Forbes, IEEE and MIT Technology Review, and Interviewed by ABC, CBS, NBC, CNN, BBC, NPR, NHK, FOX, and Washington Post. He is a member of the MIT Technology Review Global Panel. He is the author of the books: "Secure and Smart Internet of Things (IoT) using Blockchain and Artificial Intelligence (AI)" which won 3 awards San Jose State University Author and Artist Award, One of the Best Technology Books of all Time Award, and One of the Best AI Models Books of All Time Award. His second book was "Blockchain Technology and Applications" won San Jose State University Author and Artist, One of the Best New Private Blockchain Books and used at Stanford University and other prestigious schools in the USA, and "Quantum Computing" Book forthcoming in 2023. He studied Electrical Engineering at Lehigh University, Cybersecurity at Harvard University and Digital Transformation at Massachusetts Institute of Technology (MIT).

What is Blockchain?

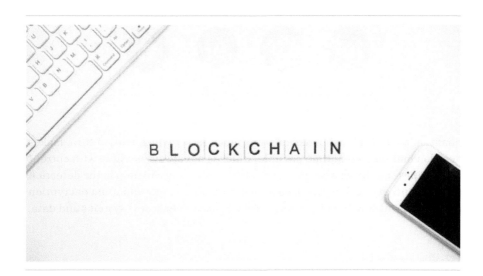

Blockchain Technology is one of the four hot technologies shaping the future of the tech world in the coming decades, these four technologies (IBAC) are: Internet of things (IoT), **B**lockchain, **a**rtificial intelligence (AI), and **c**ybersecurity (Figure 1.1). All four technologies are interconnected and impact each other in many ways. As Figure 1.2 shows, you can explain each technology with an analogy to human acts: IoT feels, Blockchain remembers, AI thinks, and

Recently, "quantum computing" presented itself as a new player impacting IBAC in many ways, for example, quantum computing will make IoT faster in processing data and extracting insights, and quantum computing will force Blockchain to invent new encryption techniques and will make processing data faster, solving one of the main issues of Blockchain technology. In the case of AI,

Figure 1.1: Simplified form of IBAC.

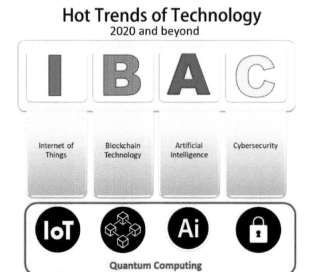

quantum computing will make analysis much faster which will, in turn, makes decision-making in real time possible, which is not always possible with current computing tools. In cybersecurity, quantum computing will help in the detection and prevention of cyber-attacks and open the door to new quantum encryption algorithms which will make it very hard for hackers to access systems and data.

1.1 What is Blockchain?

The classical definition of Blockchain is that it is "a distributed database existing on multiple computers at the same time. It constantly grows as new sets of recordings, or 'blocks', are added to it. Each block contains a timestamp and a link to the previous block, so they actually form a chain", but the best definition of Blockchain according to MIT is: *cryptography + human logic*.

If the internet is all about providing *connectivity*, Blockchain is all about enabling *trust*. For example, imagine there are 30 people in a classroom or an office building, with one main door and a security guard holding a list of authorized students/employees who can get into the building. You will need to show your card to the guard to check the list and, if you are on the list, you are in. This is the current centralized system. With the use of Blockchain, each

Figure 1.2: Hot trends of technology in 2020 and beyond.

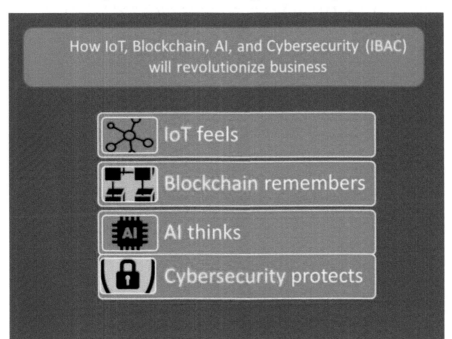

one of the 30 people will have a list with pictures of people who are authorized to be in the room so if somebody came in, and that person was not on the list, they would start talking to each other, asking "Hey, can you please check if this person belongs here?" That is synchronization and is referred to as gossip protocol within Blockchain. Human logic is the list you have, and everybody

Figure 1.3: The best definition of blockchain.

Cryptography + Human Logic = Blockchain

starting to talk to each other. On top of the current system using encryption (user name and password), we added human logic, consensus protocols, and algorithms (Figure 1.3).

1.2 The Five Components of a Blockchain

1. Cryptography
2. P2P network
3. Consensus mechanism
4. Ledger
5. Validity rules.

All listed in Figure 1.4.

Figure 1.4: The five components of a Blockchain.

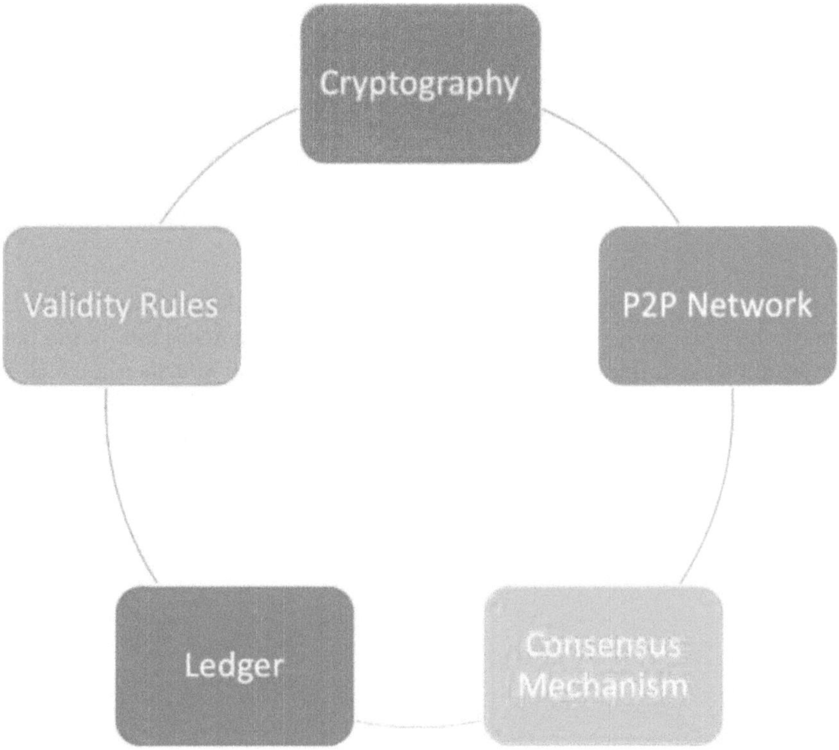

1.3 Blockchain Programming Languages

Any of the following programming languages can be used to create Blockchain platforms:

- C++ (Bitcoin)
- Python
- JavaScript
- Solidity (Smart Contract)
- Java
- Go.

Figure 1.5 shows an example of "block" programming.

Figure 1.5: An example of "block" programming.

```
class Block {
        constructor(timestamp, transactions,
        previousHash = '') {
        this.previousHash = previousHash;
        this.timestamp = timestamp;
        this.transactions = transactions;
        this.hash = this.calculateHash();
        this.nonce = 0;
        }
}
```

1.4 Mechanism of Blockchain Technology

The first block, called the genesis block, was created by a miner or validator based on consensus protocols, each block having five elements (index, time-stamp, previous hash, hash, and data). A Blockchain is initialized using the genesis block which is the foundation of the trading system and the prototype for the other blocks in the Blockchain. When you change any of these data you will change the whole block and the following blocks will see that something has changed, as will the other nodes with copies of the blocks, and the altered node will be rejected. All nodes sync using a gossip protocol – Figure 1.6 shows this type of mechanism.

A gossip protocol is a procedure or process of computer peer-to-peer communication that is based on the way epidemics spread. Some distributed systems including Blockchain use peer-to-peer gossip to ensure that data is disseminated to all members of a group. Some ad hoc networks have no central registry and the only way to spread common data is to rely on each member to pass it along to their neighbors [1].

Figure 1.6: An example of blocks of Blockchain in one node.

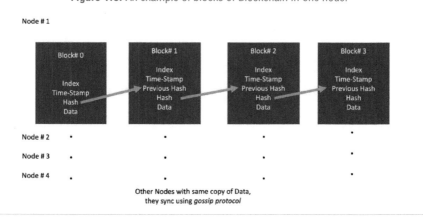

1.5 Blockchain vs. Traditional Database

There are many differences between Blockchain and traditional databases and Table 1.1 summarizes them.

1.6 The Stack of Blockchain

Like any other technology, Blockchain can be defined by its stack; Figure 1.7 explains it and it is worth emphasizing the importance of each layer as an opportunity for improvement and new business (startups), for example, UI/UX for different devices including smartphones, tablets, desktops, and laptops, in addition to the wide field of new consensus protocols for specific applications and industries, the introduction of smart contracts in the design process to avoid any surprises, and secure ways to connect the stack to the internet.

Table 1.1: Blockchain vs. a traditional database.

Blockchain vs. Traditional Database		
Characteristics	**Blockchain**	**Database**
Authority	Decentralized	Centralized and controlled by the admin
Architecture	Distributed	Client-server
Data Handling	Read and Write	CRUD (Create, Read, Update, Delete)
Integrity	High	Can be altered by hackers
Transparency	High	Controlled by the admin
Cost	High	Low
Performance	Slow	Very fast

Figure 1.7: Blockchain stack.

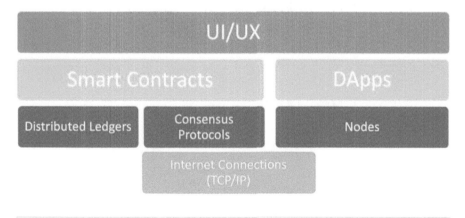

1.7 Blockchain Tracks

To understand the future direction of Blockchain technology, we need to recognize the three tracks (Figure 1.8) of Blockchain technology:

- **Pure R&D track**: This track is focused on understanding what it means to develop a Blockchain-based system. Ideally, working on real use-cases, the ultimate goal is investigation and learning, and not necessarily delivery of a working system.
- **Immediate business benefit track**: This track covers two bases: (1) learning how to work with this promising technology and (2) delivering an actual system that can be deployed in a real business context. Many of these projects are intra-company.
- **Long-term transformational potential track**: This is the track of the visionaries who recognize that to realize the true value of Blockchain-based networks means reinventing entire processes and industries as well as how public-sector organizations function.

Figure 1.8: Tracks of Blockchain technology.

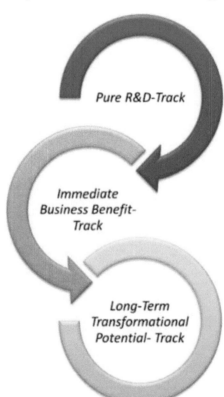

1.8 Challenges Facing Blockchain Technology

Every new technology faces challenge and Blockchain is not an exception. Both technical and non-technical challenges are listed below (Figure 1.9):

Technical Challenges

- Scalability
- Processing time
- Processing power
- 51% attack
- Double spending
- Bad smart contracts
- Storage
- First mile and last mile problem (data before and after going through the Blockchain).

Figure 1.9: Challenges facing Blockchain.

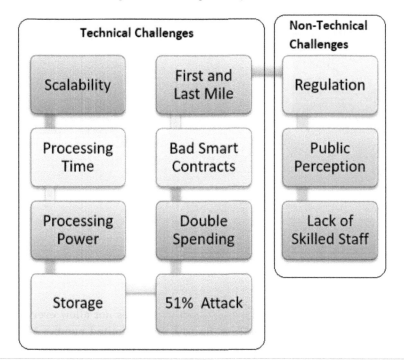

Non-technical Challenges

- Regulations
- Public perception (Blockchain is Bitcoin)
- Lack of skilled staff,

1.9 Types of Blockchain Networks

There are three types of Blockchain networks (Figure 1.10):

- **Public:** A public Blockchain is the one where everyone can see all the transactions, anyone can expect their transaction to appear on the ledger, and anyone can participate in the consensus process.

Figure 1.10: Types of Blockchain networks.

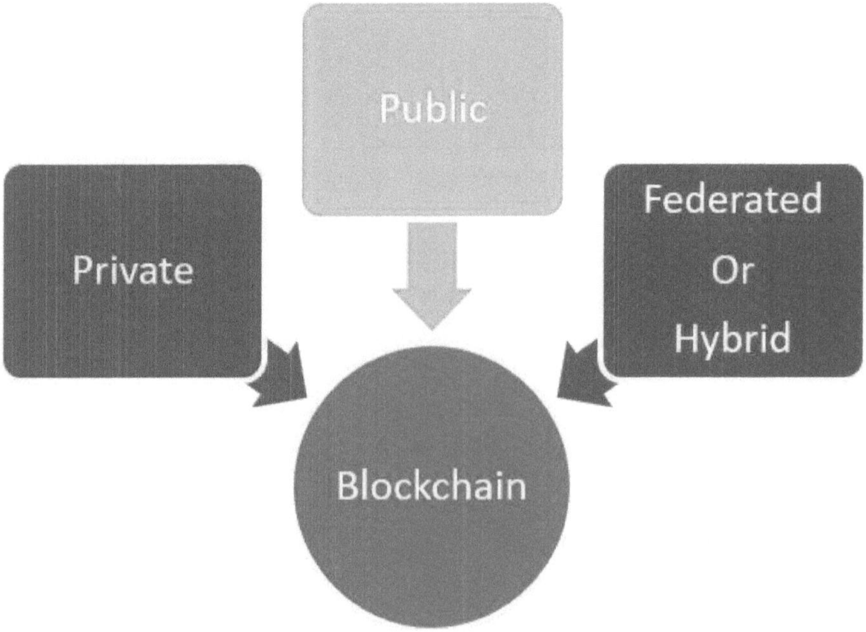

Federated/hybrid: A federated/hybrid Blockchain does not allow everyone to participate in the consensus process. Indeed, only a limited number of nodes are given permission to do so. For instance, in a group of 20 pharmaceutical companies, we could imagine that, for a block to be valid, 15 of them have to agree. Access to the Blockchain, however, can be public or restricted to the participants.

- **Private:** Private Blockchains are generally used inside a company. Only specific members are allowed to access it and carry out transactions.

2

Consensus Protocols

A consensus protocol may be defined as the mechanism through which a Blockchain network reaches consensus. Blockchains are built as distributed systems and, since they do not rely on a central authority, the distributed nodes need to agree on the validity of transactions.

This is where consensus protocols come into play. They assure that the protocol rules are being followed and guarantee that all transactions occur in a trustless way.

2.1 Types of Consensus Algorithm [2]

Below is a list of consensus algorithms. There are many other algorithms or protocols beside the ones listed here depending on the specific application and use case of Blockchain. *Proof of X*, where "X" can be any requirement, is the most exciting field of research for students and researches where creativity plays a major part in creating new con- sensus algorithms.

2.1.1 Proof of Work

Most cryptocurrencies including Bitcoin run on "proof of work". Proof of work as a process has the following steps to it:

- The miners solve cryptographic puzzles to "mine" a block to _ add to the Blockchain.
- This process requires an immense amount of energy and com- putational usage.
- The puzzles have been designed in a way that makes it hard and taxing on the system.
- When a miner solves the puzzle, they present their block to the network for verification.

Mining serves two purposes:

1. To verify the legitimacy of a transaction, or avoiding the so-called double-spending.
2. To create new digital currencies by rewarding miners for per- forming the previous task.

The miner must find a result starting with a number of zeroes. The greater the number of zeroes, the more difficult it is for the miner to find the result and the more it will have to try his luck before finding it.

Yet the number of zeroes (and therefore the difficulty) is adjusted to the number of miners on the network (and their computer capacity or hashing power) to be sure that it will take an average of 10 minutes to find the solution. Once it has found this figure, the other members of the network can instantly check the solution.

Since the miner may not find the input data ("input") from the result ("output"), they are going to try their luck until they locate the input data enabling them to obtain the output data corresponding to the objective of difficulty required, which is the number starting with a number of zeroes sufficient to be validated by the protocol Bitcoin and thus be added to the Blockchain.

Figure 2.1 is an example of the mathematical puzzle, the goal is to have three leading zeroes:

Figure 2.1: Example of a mathematical puzzle in a proof of work (PoW) protocol.

This city is amazing1 = 0ndldeouklewnlf88980378008mmkjj...

This city is amazing2 = 0ljljfdijirejopnjojnojre9980089knlkd9...

This city is amazing3 = 0uuuiiasmlmnp122339u0unnklmnkj...

.

.

.

This city is amazing409876345921 = 000jukutyghi7j6544ghjjj239...

2.1.2 Proof of Stake

Proof of stake will make the entire mining process virtual and replace miners with validators.

This is how the process will work:

1. The validators will have to lock up some of their coins as a stake.
2. After that, they will start validating the blocks. Meaning, when they discover a block they think can be added to the chain, they will validate it by placing a bet on it.
3. If the block gets appended, then the validators will get a reward proportionate to their bets.

Proof of stake is a different way to validate transactions to achieve the distributed consensus. It is still an algorithm, and the purpose is the same as the proof of work, but the process to reach the goal is quite different.

Unlike the proof of work, where the algorithm rewards miners who solve mathematical problems with the goal of validating transactions and creating new blocks, with the proof of stake, the creator of a new block is chosen in a deterministic way, depending on their wealth, also defined as stake. Also defined as stake with no block rewards.

Also, all digital currencies are created at the beginning, and their number never changes. This means that in the PoS system there is no block reward, so the miners take the transaction fees.

2.1.3 Delegated Proof of Stake (DPoS)

DPoS is similar to PoS in regard to staking but has a different and more democratic system that is said to be fair. Like PoS, token holders stake their tokens in this consensus protocol.

Instead of the probabilistic algorithm in PoS, token holders within a DPoS network are able to cast votes proportional to their stake to appoint delegates to serve on a panel of witnesses – these witnesses secure the Blockchain network. In DPoS, delegates do not need to have a large stake, but they must compete to gain the most votes from users. It provides better scalability compared to PoW and PoS as there are fully dedicated nodes who are voted to power the Blockchain. Block producers can be voted in or out at any time, and hence the threat of tarnishing their reputation and loss of income plays a major role against bad actors.

It's clear that DPoS seems to result in a semi-centralized network, but it is traded off for scalability.

2.1.4 Proof of Authority (PoA)

PoA is known to bear many similarities to PoS and DPoS, where only a group of pre-selected authorities (called validators) secure the Blockchain and are able to produce new blocks.

New blocks on the Blockchain are created only when a supermajority is reached by the validators. The identities of all validators are public and verifiable by any third party, resulting in the validator's public identity performing the role of proof of stake.

As these validators' identities are at stake, the threat of their identity being ruined incentivizes them to act in the best interest of the network. Due to the fact that PoA's trust system is predetermined, concerns have been raised that there might be a centralized element with this consensus algorithm.

However, it can be argued that semi-centralization could actually be appropriate within private/consortium Blockchains.

2.1.5 Proof of Assignment (PoA)

Similarly, to DPoS, the proof of assignment model establishes several trusted nodes within the network, but only those nodes store the entire ledger. By allowing other network contributors to participate without having ledger storage.

2.1.6 Byzantine Fault Tolerance (BFT)

BFT is the most popular permissioned (private) Blockchain platform protocol and is currently used by Hyperledger Fabric.

To understand the Byzantine fault tolerance algorithm, you need to understand the Byzantine generals problem. Imagine a group of Byzantine generals and their armies have surrounded a castle and are preparing to attack.

To win, they must attack simultaneously. But they know that there is at least one traitor among them. So, how do they launch a successful attack with at least one, unknown, bad actor in their midst?

The analogy is clear: in any distributed computing environment, Blockchain, there is a risk that rogue actors could wreak havoc. So, its reliance on community consensus makes Byzantine faults an especially thorny problem for Blockchain. PoW generally provides a solution: "Byzantine fault tolerance," but the drawbacks may not be worth it. This is where the Byzantine fault tolerance algorithm comes into play. It is considered the first practical solution to achieving a consensus that overcomes Byzantine failure. A consensus decision is determined based on the total decisions submitted by all the generals, and it addresses the challenges without the expenditure of the energy required by proof of work. However, it works only on a permissioned Blockchain because there is no anonymity.

2.1.7 Leased Proof of Stake (LPoS)

Leased proof of stake is an advanced version of the proof of stake (PoS) algorithm. Generally, in the PoS algorithm, every node holds a certain amount of cryptocurrency.

However, with leased proof of stake, users are able to lease their balance to full nodes. The higher the amount that is leased, the better the chances are that the full node will be selected to produce the next block. If the node is selected, the user will receive some part of the transaction fees that are collected by the node.

2.1.8 Proof of Elapsed Time (PoET)

Proof of elapsed time is a consensus mechanism algorithm that is often used in permissioned (private) Blockchain networks to determine the mining rights or the block winners on the network. Based on the basis of a fair lottery system where every single node is equally likely to be a winner, the PoET mechanism is based on spreading the chances of winning fairly across the largest possible number of network users.

The timer is different for every node. Every user in the network is assigned a random amount of time to wait, and the first user to finish waiting gets to commit the next block to the blockchain. Compare this to pulling straws, but this time the shortest stem in the stack wins the lottery.

2.1.9 Proof of Activity (PoA)

Proof of activity is one of the many Blockchain consensus algorithms used to ensure that all the transactions on the Blockchain are genuine and all users arrive at a consensus on the precise status of the public ledger.

Proof of activity is a mixed approach that conjoins the other two commonly used algorithms, proof of work (POW) and proof of stake (POS).

2.1.10 Proof of Importance (PoI)

Proof of importance is a consensus algorithm similar to PoS. Nodes "vest" currency to participate in the creation of blocks. Unlike PoS, proof of importance quantifies a user's support of the network.

2.1.11 Proof of Capacity (PoC)

Proof of capacity (POC) is a consensus mechanism algorithm used in Blockchains that allows the mining devices in the network to use their usable hard drive space to decide the mining rights, instead of using the mining device's computing power (as in the proof of work algorithm) or the miner's stake in the crypto coins (as in the proof of stake algorithm).

2.1.12 Proof of Burn (PoB)

Unlike PoW, proof of burn (PoB) is a consensus mechanism that does not waste energy. The real computing power is not critical to avoid manipulation. In this case, the nodes destroy or burn their tokens if they want to create the next blocks and receive a reward.

With PoB, every time a user decides to destroy some of their tokens, they buy part of the virtual computing power that gives them the ability to validate the blocks. The more tokens they burn, the higher the possibility of receiving the reward.

3

Key Blockchain Use Cases

Blockchain use cases fall into two fundamental categories: *record keeping*, static registries of data about highly valuable assets, and *transactions*, dynamic registries of the exchange of tradeable assets:

- Record keeping use cases include the long-term safeguarding of data on valuable physical and digital assets, keeping track of identity-related information about individuals and executable smart contracts based on pre-defined conditions.
- Transaction use cases include keeping track of data about frequently exchanged assets, near-real-time digital payments, and emerging digital assets.

Here are four ways that Blockchain is actually useful in avoiding pilot-to-production failure:

- The first use case is for guaranteed and verified data dissemination.
- The second use case is an asset and product tracking.
- The third use case is asset transfer.
- The fourth use case is certified claims.

Next is a comprehensive list of Blockchain applications in different industries. [3]

3.1 How Blockchain Can Help Advertising

For buy-side transparency: Blockchain for auditing; for sell-side transparency: proof of view (PoV) to fight fraud.

- PoV only records views from signed-in users, since the viewer's unique ID is part of the information required for a view to be considered valid.
- Since most people are only able to watch one video at a time, the PoV will invalidate views from a user who is streaming multiple videos simultaneously.
- The PoV technology confirms that a video is actually being streamed by capturing information about the current frame at random times.
- Smart contracts can be used to document views and who gets paid.

3.2 Verifying the Authenticity of Returned Drugs

In the pharma industry, drugs are frequently returned to the pharmaceutical manufacturers. While the proportion of the returned drugs is small compared to the sales (about 2–3% of sales), the per year volume is in the range of $7–10 billion. Currently, big pharma is working with tech companies to develop a pharma Blockchain proof of concept (PoC) app for this use case:

The system generates unique identifiers for a drug package. When a manufacturer ships a package, they register the item on the pharma PoC Blockchain, with the four pieces of information generated by the ATTP: the item number (based on GS1 standard), a serial number, a batch number, and expiration date. Using this PoC tracking will be easy, efficient, and fast.

3.3 Transparency and Traceability of Consent in Clinical Trials

Informed patient consent involves making the patient aware of each step in the clinical trial process including any possible risks posed by the study. Clinical trial consent for protocols and their revisions need to be transparent for patients and traceable for stakeholders.

However, in practice, the informed consent process is difficult to handle in a rigorous and satisfactory way. The FDA reports that almost 10% of the trials they monitor feature some issues related to consent collection. These include: failure to obtain written informed consent, unapproved forms, invalid consent documents, failure to re-consent to a revised protocol and missing institutional review board approval to protocol changes, amongst others. Frequently, there also are reported cases of document fraud such as issues of backdating consent documents.

Blockchain technology provides a mechanism for unfalsifiable time-stamping of consent forms, storing and tracking the consent in a secure and publicly verifiable way, and enabling the sharing of this information in real-time.

Additionally, smart contracts can be bound to protocol revisions, such that any change in the clinical trial protocol requires renewal of the patient consent.

3.4 Insurance

Arguably, the greatest Blockchain application for insurance is through smart contracts. Such contracts powered by Blockchain could allow customers and insurers to manage claims in a truly transparent and secure manner, according to Deloitte. All contracts and claims could be recorded on the Blockchain and validated by the network, which would eliminate invalid claims. For example, the Blockchain would reject multiple claims on the same accident.

3.5 Real Estate

The average homeowner sells his or her home every five to seven years, and the average person will move nearly 12 times during his or her lifetime. With such movement, Blockchain could certainly be of use in the real estate market. It would expedite home sales by quickly verifying finances, would reduce fraud thanks to its encryption, and would offer transparency throughout the entire selling and purchasing process.

3.6 Energy

Blockchain technology could be used to execute energy supply transactions, and it could also provide the basis for metering, billing, and clearing processes, according to PWC. Other potential applications include documenting ownership, asset management, origin guarantees, emission allowances, and renewable energy certificates.

3.7 Record Management

National, state, and local governments are responsible for maintaining individuals' records such as birth and death dates, marital status, or property transfers.

Yet managing this data can be difficult, and to this day some of these records only exist in paper form, which means sometimes citizens have to physically go to their local government offices to make changes; this is time-consuming, unnecessary, and frustrating. Blockchain technology can simplify this record keeping.

3.8 Crowdfunding

Blockchain technology, among all its benefits, can be best put to use by providing provable milestones as contingencies for giving, with smart contracts releasing funds only once milestones establish that the money is being used the way that it is said to be. By providing greater oversight into individual campaigns and reducing the amount of trust required to donate in good conscience, crowdfunding can become an even more legitimate means of funding a vast spectrum of projects and causes. Figure 3.1 shows how blockchain is revolutionizing crowdfunding. [4]

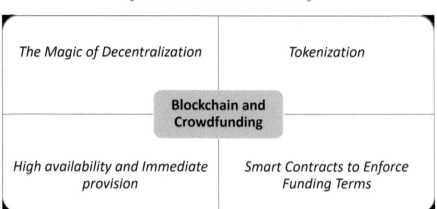

Figure 3.1: Blockchain and crowdfunding.

How Blockchain helps crowdfunding:

1. **The magic of decentralization**: Startups do not rely on any platform or combination of platforms to enable creators to raise funds. Startups are no longer be beholden to the rules, reg ulations, and whims of the most popular crowdfunding platforms on the internet. Literally, any project has a chance of getting visibility and getting funded. It also eliminates the problem of fees. While blockchain upkeep does cost a bit of money, it will cut back drastically on transaction fees. This makes crowdfunding less expensive for creators and investors. [5]

2. **Tokenization:** Instead of using crowdfunding to enable preorders of upcoming tangible products, Blockchain can use asset tokenization to provide investors with equity or some similar concept of ownership, for example initial coin offering (ICO). This way, investors will see success proportional to the eventual success of the company. This could potentially open whole new worlds of investment opportunity. Startups could save money on hiring employees by compensating them partially with fractional ownership of the business, converting it into an employee-owned enterprise. Asset tokens become their own form of currency in this model, enabling organizations to do more, like hiring professionals such as marketers and advertisers. [5]

3. **High availability and immediate provision:** Any project using a blockchain-based crowdfunding model can potentially get funded. Also, any person with an internet connection can contribute to those projects. Blockchain-based crowdfunders wouldn't have to worry about the "fraud" that has plagued modern-day crowdfunding projects. Instead contributors will immediately receive fractional enterprise or product owner- ship. [5]

4 **Smart contracts to enforce funding terms:** There are several ways in which blockchain-enabled smart contracts could provide greater accountability in crowdfunding. Primarily, these contracts would provide built-in milestones that would prevent funds from being released without provenance as to a project or campaign's legitimacy. This would prevent large sums of money from being squandered by those who are either ill-intended or not qualified to be running a crowdfunding campaign in the first place. [4]

4

Important Topics in Blockchain

There are many topics related to Blockchain that are equally important as the technology itself because they solve specific problems and challenges facing Blockchain, including forks, sharding, ZKP, and

4.1 Soft Fork vs. Hard Fork

A fork is a change to the protocol or a divergence from the previous version of the Blockchain. When a new, alternative, block is generated by a rogue miner, the system reaches consensus that this block is not valid, and this "orphan block" is very soon abandoned by the other miners.

Forks in Blockchain are of two types: *soft fork* and *hard fork*.

4.1.1 Soft Fork

A soft fork is a software upgrade that is *backward compatible with older versions.* This means that participants that did not upgrade to the new software will still be able to participate in validating and verifying transactions.

It is much easier to implement a soft fork as only a majority of participants need to upgrade the software. All participants, whether they have updated or not will continue to recognize new blocks and maintain compatibility with the network. A point to take note, however, is that the functionality of a non-upgraded participant is affected. An example of a soft fork is when the new rule states that the block size will be changed from the current 1MB (1000 KB) to 800 KB. Non-upgraded participants will still continue to see that the incoming new transactions are valid. The issue is that, when non-upgraded miners try to mine new blocks, their blocks (and thus, efforts) will be rejected by the network. Hence, soft forks represent a gradual upgrading mechanism as those who have yet to upgrade their software are incentivized to do so, or risk having reduced functionality.

4.1.2 Hard Fork

Hard forks refer to a software upgrade that is not compatible with older versions. All participants must upgrade to the new software to continue participating and validating new transactions. Those who did not upgrade would be separated from the network and would not be able to validate the new transactions. This separation results in a permanent divergence of the Blockchain.

As long as there is support in the minority chain, in the form of participants mining in the chain, the two chains will concurrently exist. Examples: Ethereum Classic and Bitcoin Cash.

4.1.3 Ethereum Classic

Ethereum had a hard fork to reverse the effects of a hack that occurred in one of their applications (called the Decentralized Autonomous Organization or simply, DAO). However, a minority portion of the community was philosophically opposed to changing the Blockchain at any cost, to preserve its nature of immutability.

As Ethereum's core developers and the majority of its community went ahead with the hard fork, the minority that stayed behind and did not upgrade their software continued to mine what is now known as Ethereum Classic (ETC).

It is important to note that since the majority transited to the new chain, they still retained the original ETH symbol, while the minority supporting the old chain were given the term Ethereum Classic or ETC.

4.1.4 Bitcoin Cash

Bitcoin Cash is a cryptocurrency that is a fork of Bitcoin. Bitcoin Cash is a spin-off or altcoin that was created in 2017. In 2018, Bitcoin Cash subsequently split into two cryptocurrencies: Bitcoin Cash and Bitcoin. Bitcoin Cash is sometimes also referred to as Bcash. [6]

Bitcoin was forked to create Bitcoin Cash because the developers of Bitcoin wanted to make some important changes to Bitcoin. The developers of the Bitcoin community could not come to an agreement concerning some of the changes that they wanted to make, so a small group of these developers forked Bitcoin to create a new version of the same code with a few modifications. [7]

The changes that define the difference between Bitcoin Cash and Bitcoin are these (Table 4.1):

- Bitcoin Cash has cheaper transfer fees, so making transactions in BCH will save you money against using BTC.
- BCH has faster transfer times, so you do not have to wait the 10 minutes it takes to verify a Bitcoin transaction!
- BCH can handle more transactions per second. This means that more people can use BCH at the same time than with BTC.

Table 4.1 compares both cryptocurrencies. [8]

4.2 Zero-knowledge Proof (ZKP)

In cryptography, zero-knowledge proof or a zero-knowledge protocol is a method by which one party (the prover) can prove to another party (the verifier) that they know a value x, without conveying any information apart from the fact that

Table 4.1: Bitcoin vs. Bitcoin Cash.

Bitcoin	Bitcoin Cash
Standard Block Size: 1MB Max	PowerBlocks: 8 MB Max
Transactions Signatures can be dis- carded from the Blockchain	All transactions signatures must be validated and secured on the Blockchain
Single Centralized Development Team and Client Implementation	Multiple Independent Development Teams and client implementations

they know the value of x. Zero-knowledge proof is the ability to prove a secret without revealing what the secret is.

4.2.1 How Does Zero-knowledge Proof Work?

The best way to explain the process of zero-knowledge proofs is with a non-digital example, which is, of course, far from the complexity of zero-knowledge proofs but very well explains how they work.

Scenario 1

Let us assume there is a blind person who has two balls, one black and one white. You then would like to prove to the blind person that these balls are indeed of different colors without revealing the color of each ball. For this, you ask the blind person to hide both balls under the table and bring one ball back up for you to see. After that, he should hide the ball back under the table and then either show the same ball or the other one. As a result, you can prove to the blind person that the colors are different by saying whether he changed the balls under the table or not.

Scenario 2

Pretend that there's a circular cave, with only one entrance or exit and at the back of this circular cave there's a door that can be unlocked using a secret code entered onto a keypad (Figure 4.1). If I want to prove to you that I know the unlock code without revealing that unlock code to you, all I need to show is that I can walk into one end of the cave, open the door, and come out the other end.

Figure 4.1: Cave example of ZKP.

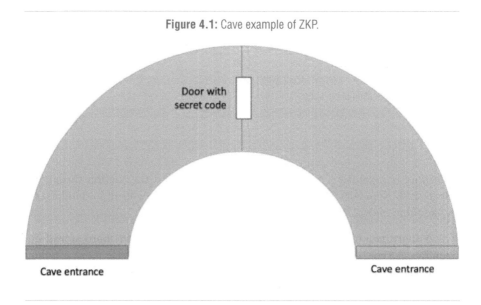

Door with
secret code

Cave entrance   Cave entrance

If I have successfully demonstrated that, then you know without a doubt I have been able to unlock that door, but yet I have not revealed that unlock code to you.

4.2.2 Zero-knowledge Proofs in Blockchain

Zero-knowledge protocols enable the transfer of assets across a distributed, peer-to-peer Blockchain network with complete privacy. In regular Blockchain transactions, when an asset is sent from one party to another, the details of that transaction are visible to every other party in the network.

By contrast, in a zero-knowledge transaction, the others only know that a valid transaction has taken place, but nothing about the sender, recipient, asset class, and quantity. The identity and amount being spent can remain hidden. For example, a user may make a request to send another user some money. The Blockchain naturally wants to make sure, before it commits this transaction, that the user who sends the money has enough money to send. However, the Blockchain doesn't really need to know or care who is spending the money, or how much total money they have. Hyperledger Fabric and Ethereum have implementation of zero-knowledge proofs under development.

4.2.3 ZKP Advantages

1. Zero-knowledge transfer, as the name suggests.
2. Computational efficiency: no encryption.
3. No degradation of the protocol.

4.3 Sharding

Sharding is a solution to the scalability, latency, and transaction throughput issues in Blockchain. Sharding is a concept that is widely used in databases to make them more efficient.

A shard is a horizontal portion of a database, with each shard stored in a separate server. This spreads the load and makes the database more efficient. In the case of the Blockchain, each node will have only a part of the data on the Blockchain, and not the entire information, when sharding is implemented.

Nodes that maintain a shard maintain information only on that shard in a shared manner, so within a shard the decentralization is still maintained. However, each node does not load the information onto the entire Blockchain, thus helping scalability. Blockchains that implement sharding use a proof of stake (PoS) consensus algorithm.

4.4 What is Sharding in Blockchain?

In order to tackle the currently persisting issues with validation mechanisms, a new kind of validation protocol has been created, i.e., sharding. As part of sharding, only a small subset of nodes (called a *shard*) out of the entire network nodes will carry out validation of every single transaction. A sharding schema is comprised of a cluster isolated shards.

The Ethereum network will be logically divided into multiple shards in the same way that a country is divided into multiple states in order to have a better governance system. Transactions created by users or a particular Shard will be validated by miner's percentage in that shard alone.

4.4.1 Drawbacks of Sharding

If you think sharding is the holy grail of all the scaling and performance issues, then you are mistaken. Sharding does come with its share of issues. The biggest flaw with sharding is that inter-shard communication is not very easy. What this basically means is that, as long as communication occurs with a shard, the picture remains rosy and nice. But if a user (e.g. Bob) who belongs to shard-1 wants to transact with another user (say John) from shard-2, the transaction would require some special protocols to complete the transaction.

The developer community is the most affected as they will have to program their codes to handle this.

4.5 What Is a Smart Contract?

A smart contract is a self-executing contract with the terms of the agreement between buyer and seller being directly written into lines of code. The code and the agreements contained therein exist across a distributed, decentralized Blockchain network. The code controls the execution, and transactions are trackable and irreversible. Smart contracts permit trusted transactions and agreements to be carried out among disparate, anonymous parties without the need for a central authority, legal system, or external enforcement mechanism. [9]

Smart contracts can be termed as the most utilized application of Blockchain technology in current times. The concept of smart contracts was introduced by Nick Szabo, a legal scholar and cryptographer, in 1994. He came to the conclusion that any decentralized ledger can be used as self-executable contracts which, later on, were termed smart contracts. These digital contracts could be converted into code and allowed to run on a Blockchain.

Smart contracts are one of the most successful applications of Blockchain technology. Using smart contracts in place of traditional ones can reduce the transaction costs significantly. Ethereum is the most popular Blockchain platform for creating smart contracts. It supports a feature called *Turing completeness* that allows the creation of more customized smart contracts. Smart contracts can be applied in different industries and fields such as smart homes, e-commerce, real estate, asset management, etc. [10]

Smart contracts are automatically executable lines of code that are stored on a Blockchain which contain predetermined rules (Figure 4.2) [11]. When these rules are met, these codes execute themselves and provide the output. In the simplest form, smart contracts are programs that run according to the format that they've been set up in by their creator. Smart contracts are most beneficial in business collaborations in which they are used to agree upon the decided terms set up by the consent of both the parties. This reduces the risk of fraud and as there is no third-party involved and the costs are reduced too. To summarize, smart contracts usually work on a mechanism that involves digital assets along with multiple parties where the involved participants can automatically govern their assets. These assets and be deposited and redistributed among the participants according to the rules of the contract. Smart contracts have the potential to track real-time performance and save costs.

Smart contracts properties:

- Self-verifiable
- Self-executable
- Tamper proof.

Figure 4.2: Example of smart contract code using Solidity (storage contract).

```solidity
pragma solidity >=0.4.0 <0.7.0;

contract SimpleStorage {

    unit storedData;

    function set(unit x) public {

        storedData = x;

    }

    function get() public view returns (unit) {

        return storedData;
    }

}
```

5

Decentralized Applications – DApps

Decentralized applications (DApps) are applications that run on a P2P (peer-to-peer) network of computers rather than a single computer. DApps have existed since the advent of P2P networks. They are a type of software program designed to exist on the Internet in a way that is not controlled by any single entity. As opposed to simple smart cotracts, in the classic sense of Bitcoin, which sends money from A to B, DApps have an *unlimited* number of participants on all sides of the market.

5.1 Difference Between DApps and Smart Contracts

DApps are a "Blockchain-enabled" website, where the smart contract is what allows it to connect to the Blockchain. The easiest way to understand this is to understand how traditional websites operate. The traditional web application uses HTML, CSS, and JavaScript to render a page. It will also need to grab details from a database utilizing an API. When you go onto websites like Facebook, the page will call an API to grab your personal data and display them on the page (Figure 5.1).

Figure 5.1: Traditional website process.

DApps are similar to a conventional web application. The front end uses the *exact same* technology to render the page. The one critical difference is that instead of an API connecting to a Database, you have a smart contract connecting to a Blockchain (Figure 5.2).

Figure 5.2: DApp enabled website.

As opposed to traditional, centralized applications, where the backend code is running on centralized servers, DApps have their backend code running on a decentralized P2P network. Decentralized applications consist of the whole package, from backend to frontend, but the smart contract is only one part of the DApp:

- Frontend (what you can see) and backend (the logic in the background).
- A smart contract, on the other hand, consists only of the backend, and often only a small part of the whole DApp.

This means that if you want to create a decentralized application on a smart contract system, you have to combine several smart contracts and rely on third party systems for the frontend.

DApps can have frontend code and user interfaces written in any language (just like an App) that can make calls to its backend. Furthermore, its frontend can be hosted on decentralized storage.

5.2 Blockchain DApps

For an application to be considered a DApp in the context of Blockchain, it must meet the following criteria:

The Application Must be Completely Open-source

It must operate autonomously, and with no entity controlling the majority of its tokens. The application may adapt its protocol in response to proposed improvements and market feedback, but a consensus of its users must decide all changes.

The Application's Data and Records of Operation Must be Cryptographically Stored

They must be cryptographically stored in a public, decentralized Blockchain in order to avoid any central points of failure.

The Application Must use a Cryptographic Token

This is a Bitcoin or a token native to its system that is necessary for access to the application and any contribution of value from miners should be rewarded with the application's tokens.

The Application Must Generate Tokens

According to a standard cryptographic algorithm acting as a proof of the value, nodes are contributing to the application (e.g. Bitcoin uses the proof of work algorithm).

5.3 Example: Ethereum DApps

Ethereum provides developers with a foundational layer: a Blockchain with a built-in Turing-complete programming language, allowing anyone to write smart contracts and decentralized applications where they can create their own arbitrary rules for ownership, transaction formats, and state transition functions.

In general, there are three types of applications on top of Ethereum:

Financial Applications

Providing users with more powerful ways of managing and entering into contracts using their money.

Semi-financial Applications

Where money is involved, but there is also a heavy non-monetary side to what is being done

Governance Applications

Such as online voting and decentralized governance that are not financial at all.

References

[1]   https://en.wikipedia.org/wiki/Gossip_protocol

[2]   https://medium.com/the-daily-bit/9-types-of-consensus-mecha-nisms-that-you-didnt-know-about-49ec365179da

[3]   https://www.businessinsider.com/blockchain-technology-applica-tions-use-cases-2017-9

[4]   https://www.disruptordaily.com/blockchain-use-cases-crowdfund-ing/

[5]   https://due.com/blog/a-new-era-of-crowdfunding-blockchain/

[6]   https://en.wikipedia.org/wiki/Bitcoin_Cash

[7]   https://www.bitdegree.org/tutorials/bitcoin-cash-vs-bitcoin/

[8]   https://btcmanager.com/us-authorities-blockchain-covid-19-critical-services/?q=/us-authorities-block-chain-covid-19-critical-services/&

[9]   https://www.investopedia.com/terms/s/smart-contracts.asp

[10] https://hackernoon.com/everything-you-need-to-know-about-smart-contracts-a-beginners-guide-c13cc138378a

Index